Nisqually Watershed

Nisqually Watershed

GLACIER TO DELTA ❖ A RIVER'S LEGACY

TEXT BY DAVID GEORGE GORDON ❖ PHOTOGRAPHY BY MARK R. LEMBERSKY

THE MOUNTAINEERS ❖ NISQUALLY RIVER INTERPRETIVE CENTER FOUNDATION

Published by The Mountaineers
1011 SW Klickitat Way, Seattle WA 98134
(All trade inquiries to this address)
and
Nisqually River Interpretive Center Foundation,
P.O. Box 759, Yelm WA 98597

Cover/book design and typography: Elizabeth Watson
Illustrations: Jim Hays
Map: Vikki Leib

PHOTOS. *Front cover and title page:* Nisqually River vista, middle watershed. *Page 5:* Alder Lake shore.
Page 6: Mount Rainier summit. *Pages 8 and 9:* Tule Lake pond lilies. *Page 128 and back cover:* From glacier to delta.

HISTORICAL PHOTO CREDITS. *Pages 44, 51:* Special Collections Division, University of Washington Libraries.
Pages 47, 53, 53 top, 98, 109: Washington State Historical Society, Tacoma, WA. *Page 49, top:* D. Kinsey Collection,
Whatcom Museum of History and Art, Bellingham, WA. *Page 49 bottom:* Special Collections, Tacoma Public Library.
Page 55: National Park Service. *Pages 57, 59:* Tacoma Public Utilities. *Page 70:* James Daley. *Page 77:* Bert Kellogg Collection
of the North Olympic Library System. *Page 79:* Charles L. Pack Experimental Forest. *Page 81:* State University of
New York College of Environmental Science and Forestry. *Page 86:* Courtesy of David Peissner.
PHOTO CREDIT. *Page 11, 13:* Courtesy of Northwest Indian Fisheries Commission.

Library of Congress Cataloging-in-Publication Data
Gordon, David G. (David George), 1950–
      Nisqually watershed : glacier to delta - a river's legacy / text by David George Gordon / photography by
  Mark R. Lembersky
        p. cm.
     Includes index.
     ISBN 0-89886-453-4
     1. Ecosystem management—Washington (State)—Nisqually River Watershed. 2. Man—Influence on nature—
  Washington (State)—Nisqually River Watershed. 3. Natural history—Washington (State)—Nisqually River
  Watershed. 4. Nisqually River Watershed (Wash.)—History. I. Lembersky, Mark R. II. Title.
QH76.5.W2G67 1995
508.797 '78—dc20   95-11067
                                            CIP

Printed on acid-free paper in Hong Kong.

Acknowledgments

The Nisqually River Interpretive Center Foundation gratefully acknowledges the generous financial assistance for this book from:

Key Bank of Washington
Mrs. Yvonne McElroy
Tosco Northwest Company/BP
Weyerhaeuser Company Foundation
Wilcox Family Farm.

The Foundation appreciates the involvement and encouragement of the Nisqually River Council and its member organizations: Pierce County; Thurston County; Lewis County; State of Washington Department of Natural Resources, Department of Fish and Wildlife, Department of Ecology, Department of Agriculture, Parks and Recreation Commission and Secretary of State; Nisqually National Wildlife Refuge; Mount Rainier National Park; U.S. Army at Fort Lewis; Gifford Pinchot National Forest; Nisqually Indian Tribe; University of Washington Pack Experimental Forest; Tacoma Public Utilities; Municipalities of Yelm, Eatonville and Roy; and Nisqually River Citizens Advisory Committee.

A manuscript, like a river, is fed from many sources. This book draws on information, both published and unpublished, from several watershed scholars, including Cecelia Svinth Carpenter, Carolyn L. Dreidger, Ruth Kirk, Peter Moulton, Gene Allen Nadeau and George Walter. David George Gordon thanks these individuals for their generosity and guidance. Thanks also to the many residents of the watershed who shared their knowledge and showed their hospitality during his frequent field trips to the Nisqually River. Special thanks are due to Nora Deans, Margaret Foster, Donna DeShazo, Rachel Bard, Lela Hilton and Laura Popenoe for their roles in shaping the text and to Rae Nelson for compiling the book's index.

Mark Lembersky is indebted to the many people who helped make possible the photographs in this book. These images evolved from a meeting attended with Pat Walter, an invitation from Dan Treat to photograph the confluence of the Nisqually and Mashel rivers, and encouragement from Steve Craig and Peter Moulton to visually record the watershed. Numerous individuals and organizations generously provided access by land, air and water to sites seen on these pages. They include the U.S. Army at Fort Lewis, National Fish and Oyster Company, Nisqually Indian Tribe, Pioneer Farm Museum, Weyerhaeuser Company and Wilcox Family Farm. Rae Nelson offered many insightful suggestions, and Elizabeth Watson artfully designed the book.

ACKNOWLEDGMENTS   7

FOREWORD   11

WATERSHED   *The Gathering Place*   17

ALCHEMY   *Origins of the Upper Watershed*   37

CONFLUENCE *Meetings in the Middle Watershed*   65

DELTA   *Life in the Lower Watershed*   91

PUBLISHERS' NOTES   123

ADDITIONAL INFORMATION   125

INDEX   126

Foreword

For thousands of years, Nisqually Indians have lived along the river that bears our name. My father lived with this river for 104 of those years, nourished by its natural resources and spiritual gifts. He was born in a cedar longhouse overlooking the Nisqually and was raised to respect and appreciate the many values of our homeland.

In his lifetime, Dad saw many changes to the river and the land just beyond its banks. During his life, he saw the decline of the great salmon runs, the depletion of wildlife and, in places such as the prairies, the disappearance of

◄ ◄ *Sandbar at mid-river.*

◄ *Bill Frank, Jr., with his father.*

11

native plants. Still, when my dad recounted boyhood memories of harvesting the wild plants or traveling with his family by dugout canoe, you could see his eyes sparkle with reflections of a paradise gone by.

Dad knew, and he taught me, just as other Nisqually parents and grandparents taught their children, that our Nisqually homeland is a unique and beautiful place. When he died in 1983, he left a legacy that will not be forgotten—his vision of protecting the river, its tributaries and land that drains to these waters. From him I learned how it is our job to conserve the resources of the Nisqually and to continue our fishing way of life. This is our Nisqually cultural heritage, and it will last beyond our lifetimes to those of our children— and to their children. The health of our Nisqually community depends on the health of the Nisqually River.

This is not to say that people cannot use the productive natural resources of the Nisqually River Watershed. There is room for the farmer, rancher and forester, as well as for the Indian fisherman. However, those who use these natural resources also must accept the responsibility for good stewardship. Decisions made today must insure a healthy and productive natural resource base for future generations.

In recent years something very encouraging has been happening to provide protection, both now and for the future, for the Nisqually and other watersheds in America. Ordinary citizens are working together with government and private interests to plan for the conservation of our natural resources. In Washington State, people from all walks of life helped produce the Nisqually River Management Plan, a blueprint for stewardship of the watershed, based on cooperation, that will make sure that the river will flow strong and clean.

One of my great concerns today is how we meet the challenge of passing to our children this message—my dad's message—of responsible stewardship. This book, with its story in pictures and words, is an important step in recording the legacy of the Nisqually River Watershed—its history, economies, environment and culture— and passing it to others, young and old.

It is the kind of story that my dad might have told, full of colorful characters, rich scenes from nature and many important lessons for us all to learn. My eyes sparkle, as Dad's did, whenever I connect with that place in my heart where the Nisqually flows, timeless and sure. I hope that through this book, you, too, will connect with that place in your heart.

Bill Frank, Jr.
Tribal Elder and recipient of the
United Nations' Albert Schweitzer Humanitarian Award

Graham

5

161

Sequalitchew Cr.

Puget Sound

Dupont

7

507

Fort Nisqually Sites

Nisqually Reach

Nisqually Flats

Nisqually R. Delta

Muck Cr.

FORT LEWIS MILITARY RESERVATION

13th Division Prairie

Tanw... L...

South Cr.

Brown Farm Site

Nisqually Nat'l Wildlife Refuge

Clear L...

Nisqually River

Clear Cr.

Nisqually Lake

NORTHWEST WILDLIFE

Nisqually

Roy

McAllister Creek

McAllister Springs

NISQUALLY INDIAN RESERVATION

Muck Cr.

702

Tanwax Cr.

510

Lake St. Clair

McKenna

Nisqually River

Silver Lake

Ohop V...

Eaton Cr.

Thompson Cr.

Harts Lake

Nisqu... Ma... State ...

Centralia Canal

Diversion Dam

Kreger Lake

Ohop Cr.

Yelm

Tule Lake

N

E

W

S

Yelm Cr.

Weir Prairie

Rainier

Bald Hill Natural Area Preserve

Elbow Lake

Bald Hill Lake

Clear Lake

Vail

Nisqually Watershed

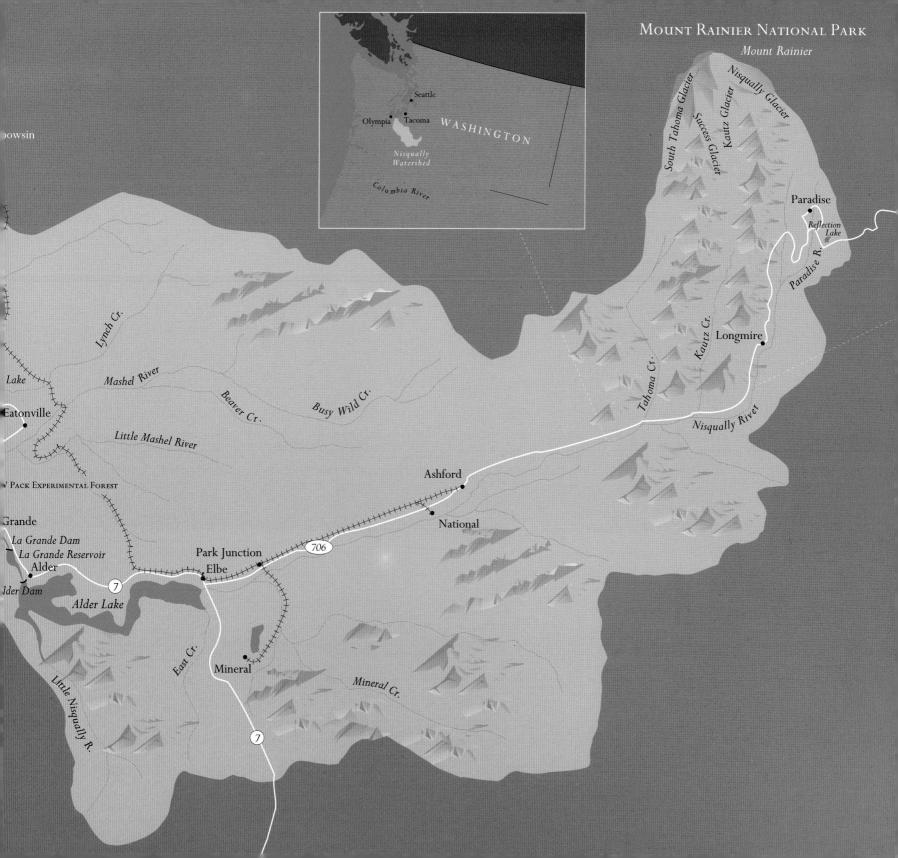

MOUNT RAINIER NATIONAL PARK

Mount Rainier

WASHINGTON

Seattle

Olympia Tacoma

Nisqually Watershed

Columbia River

South Tahoma Glacier

Success Glacier

Kautz Glacier

Nisqually Glacier

Paradise

Reflection Lake

Paradise R.

Longmire

Tahoma Cr.

Kautz Cr.

Nisqually River

owsin

Lynch Cr.

Lake

Mashel River

Eatonville

Little Mashel River

Beaver Cr.

Busy Wild Cr.

W Pack Experimental Forest

Ashford

National

Grande

La Grande Dam

La Grande Reservoir

Alder

Park Junction

706

lder Dam

Elbe

Alder Lake

East Cr.

Mineral

Little Nisqually R.

Mineral Cr.

7

7

WATERSHED

WATERSHED

The Gathering Place

The most prominent feature of the Nisqually River Watershed, Mount Rainier is a comparatively recent addition to the landscape. This snow-covered volcano, which rises in splendid isolation more than 14,000 feet above its surroundings, is perhaps no more than a million years old. It rests on layers of sandstone and basalt that date back 40 million years, to a time when the coastline of the Pacific Northwest was newly formed.

However young, the mountain has made its mark on the landscape of Washington State. Its frequent eruptions, some as recent as 150 years ago, tossed spongy, lightweight pumice and hot volcanic rock "bombs" across the land. Molten lava flowed from its cone on many occasions, cooling and solidifying into formations of granite and basalt rock. Mudslides and debris-flows filled the valleys and plains at its foot—in one instance covering a hundred square miles and burying portions of the land under 70 feet of clay and volcanic ash.

The mountain also influences the weather. A natural barrier to winds blowing inland from the Pacific Ocean, its cone causes air currents to rise and become cooler, releasing their moisture in the form of rain and snow. Most of this rain falls on the west slopes of the mountain and its foothills. Near the mountain's top, this abundant precipitation collects in deep layers of snow.

◄ ◄ (PREVIOUS PAGE)

Water in the clouds,

atop Mount Rainier.

◄ Falling snow

decorates a bridge

over the Paradise

River, upstream of

Narada Falls.

▶ Rock, dust and debris peek

out from the glacier's topmost layer of

snow during summer.

▶ ▶ The young river follows a

slalom-like course.

Compressed by the weight of successive layers over thousands of years, the snow has become packed into dense formations of ice, or glaciers. One of these, the Nisqually Glacier, is more than 400 feet deep in places and covers nearly two square miles of mountainside.

Buried deep beneath shards of falling rock and dust from the volcano's cone, the Nisqually Glacier's lower third can be difficult to distinguish from the surrounding moraines. Its dusty covering draws heat from the sun and, where this layer is less than six inches thick, transmits warmth to the snowpack below.

Solid becomes liquid—glittering, jewel-like water droplets that ooze from the glacier's frozen walls. Some of these shimmering beads collect in small pools on the surface. Others course through a network of tubes and channels in the glacier's interior.

From these droplets atop Mount Rainier, the Nisqually River's 78-mile-long journey begins, a circuitous route to the sea, full of cascades and riffles, switchbacks, oxbows and bends. For fellow travelers on this excursion, the trip is like no other in the world.

The Nisqually River is the only body of water in the United States with its headwaters in a national park and its mouth in a national wildlife refuge. It is one of a handful of rivers in western Washington still bordered on both sides by large tracts of coniferous forest and open expanses of prairie lands. Less than an hour's drive from two burgeoning metropolitan centers and spanned by Interstate 5, the Northwest's most heavily traveled superhighway, the river's setting somehow remains pastoral—a patchwork of farm fields and pastures, public parks and open spaces, placid ponds and shimmering lakes.

The river's apparent health is no accident. Neither are the rich timber lands, from which millions of board feet have been logged, nor the healthy runs of salmon that return to the Nisqually and its tributaries to spawn every year. These and other signs of vitality can be attributed in significant measure to the concerted stewardship of land and water in recent times.

The splendor here arises from more than the majestic mountain, magic river or fertile meeting place with Puget Sound. It is also the commitment to the well-being of the Nisqually River, and the Watershed through which it flows, that makes this 720-square-mile rectangle of public and privately owned land unique.

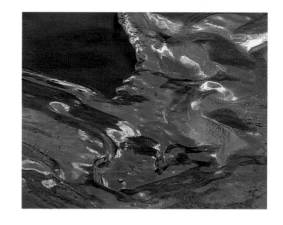

◄ AND ▲ *Icy but oxygen-rich water fills the Nisqually River, in the upper Watershed.*

► (FOLLOWING PAGE) *Morning sun cuts through the mist, spotlighting the forested banks of the middle Watershed.*

WHAT IS A WATERSHED?

The word watershed means "a gathering place, a collecting of water," and is often used to describe a region, big or small, in which an amassing of moisture occurs. Within a watershed, snow, rain, rivers, streams, lakes, ponds, wetlands and groundwater aquifers are all links in an intricate chain called the hydrologic cycle. In this cycle, rain falls on the land and soaks into the ground. Some of the water evaporates, some is absorbed by trees, shrubs, grasses and ground covers and some flows across the land to feed creeks and streams.

Water that soaks into the ground follows a maze of cracks in the bedrock, replenishing groundwater supplies. Slowly metered from subterranean store-houses, this groundwater nurtures streams, wetlands and, in most watersheds, people whose drinking water comes from shallow springs and deep wells.

Much of the water that runs across the land finds its way to progressively larger bodies of water—from creeks to rivers, rivers to bays and bays to seas. In

Antlers—crowning glory of a male elk. Stags shed them in early spring, then grow new ones in summertime.

25

western Washington, most surface water empties into Puget Sound, a vast mixing bowl for fresh and salt water supplies that was deeply scoured by glacial activity during the late Pleistocene, some 15,000 years ago. An estimated 140 billion cubic feet of fresh water pours into the Sound each year. Half of the volume that enters southern Puget Sound flows from the Nisqually River.

Carried to the Pacific Ocean by tidal currents and winds, much of this water evaporates, rising skyward to form clouds. Drifting inland, these clouds eventually release their moisture as snow and rain—and the hydrologic cycle continues, as it has for many millions of years.

Every part of a watershed is linked by the hydrologic cycle, so every change, no matter how small or remote, has the capacity to affect everything else. In nature, these changes may be as quiet as the twigs and branches that collect behind a fallen log, gradually altering the path of a stream. Or they may be as dramatic as winter floods that sweep away houses, property and even people in their paths.

PLANS FOR PROTECTION

Whether subtle or bold, any alterations we make to the hydrologic cycle can have profound consequences. Soil that erodes from a construction site in the upper part of a watershed may wash into shellfish beds at a river's mouth many miles downstream. Here, it chokes the oysters and clams that need silt-free water to survive.

Downstream actions also affect life upstream. If marshlands at a river's mouth are converted to farm fields, waterfowl and other animals may lose feeding and breeding grounds and disappear from sites throughout the watershed's upriver habitats.

▶ The river wears many faces, some turbulent and others placid, as it flows to the sea.

People have been altering the Nisqually River Watershed environment since their arrival thousands of years ago. At first, the alterations were comparatively slight—the selective harvest of cedar trees for canoes and longhouses, or the placing of fish traps and weirs in salmon-bearing streams. However, as more people moved into the Watershed, they brought along large-scale farming, forestry and hydroelectric technologies, which amplified their effects on the Watershed. Over the last 50 years, the potential for causing harm to the Nisqually has grown substantially.

Rapid population growth in cities beyond the Watershed's boundaries, combined with the ever-increasing appeal of rural lifestyles, places a sizable burden on area resources, potentially upsetting the delicate balance of water and land. Fueled by rapid regional growth, the market for rural real estate could boom, causing the conversion of many farmlands and small woodlots to sites for vacation cabins and year-round homes.

Population growth within the Watershed poses equal threats to water quality and wildlife habitats. Rapid development without adequate plans for shoreline protection, septic system installation and maintenance, stormwater runoff control and other environmental needs could prove disastrous if allowed to run its course. Several once-rural communities in the Watershed are now exploring ways to address these issues without infringing on individual property rights.

These challenges are hardly unique to the Nisqually River Watershed. Nearly all watersheds in the Puget Sound area and many other locales in the Pacific Northwest face similar problems. It is no coincidence that the 1987 Puget Sound Water Quality Management Plan called for watershed planning region-wide, directing local governments to formulate pollution control plans for the

▶ The river's path has shifted many times in the past, and, as water supplies and neighboring land uses change, it is certain to shift again.

top-ranked watersheds in each of 12 Puget Sound counties. Over three dozen plans have been drafted for Puget Sound watersheds, with dozens more in development.

WATERSHED STAKEHOLDERS

Efforts to conserve the Nisqually River Watershed have not been motivated by crisis nor steered by confrontation. They have arisen from the desires of public and private stakeholders to conserve the unique characteristics of this valued region.

Recognizing the importance of the river and the broadbased support for measures to ensure its conservation, the Washington State Legislature voted in 1985 to "initiate a process that emphasizes the natural and economic values . . . and that will bring about a stewardship program for the Nisqually River." The overall goal of such a program, the Legislature proclaimed, was to "assure enhancement of economic and recreational benefits for this generation as well as those to come."

Their first step was to form the Nisqually River Task Force, a regional planning body that brought together representatives from public resource management agencies, private citizens' groups, the Nisqually Indian Tribe, individual land owners and timber, agriculture and hydropower interests. Task Force members first reviewed information about the Watershed, then drafted a set of policy recommendations, which became a single 19-page document—the Nisqually River Management Plan. Adopted by the Legislature in 1987, the plan addresses issues of public access to the river, natural resource protection and enhancement, flood control and the status of rural landscapes and economies.

◀ Summer rain, a sylvan stretch of the Mashel River.

31

Drafting a management plan for the Nisqually River was one thing; carrying out such a plan in a smooth and effective process was another. To assist in its implementation, the Nisqually River Council was formed, representing citizen interests and federal, state, local and tribal government. The Council is an advisory group committed to the protection and enhancement of the river and its surrounding lands through education, advocacy and coordination. A separate Citizens Advisory Committee made up primarily of Watershed residents helps to fine-tune various facets of the Nisqually River Management Plan.

The original plan emphasized mostly shoreline areas—a corridor one-half to three-quarters of a mile wide on either side of the river and its tributaries. However, the Council soon enlarged the plan's focus to embrace the entire Nisqually River Watershed, a measure that recognized the interconnectedness of rivers, streams, wetlands, ponds and lakes as well as the upland areas that contribute water to these bodies.

Decision makers throughout Washington and other parts of the nation are following the Council's lead in pursuing watershed-based planning. The synergistic strengths of the Nisqually River Watershed management program were formally recognized with a 1992 Environmental Excellence Award from the state and a regional Administrator's Award from the U.S. Environmental Protection Agency. The Nisqually itself has become a model watershed, its progress closely watched by the President's Council on Sustainability and such entities as the World Resources Institute, Friends of the Earth and the governments of Canada and Japan.

Within a watershed, we all live downstream. This message from the Nisqually River now echoes loudly in watersheds across North America and around the world.

▶ *A placid moment on the Nisqually Flats, at the end of the river's 78-mile downhill journey.*

Watershed Words:
A Language of Its Own

THE TERMS FOR describing a river and its watershed are a mix of old and new. For example, people have been using the word *rill,* a very small brook, for many centuries. On the other hand, *wetland*—a word that encompasses swamps, bogs, estuaries and other areas of high soil moisture—is considerably more recent and has yet to be fully incorporated into our day-to-day conversations.

Webster's Dictionary defines a watershed as "a region or an area bounded by a water parting and draining ultimately to a particular watercourse or body of water." Other words for a watershed's features—some familiar, others foreign to the ear—should also be defined:

Habitat is the living place of fish, wildlife and plants. Clean gravel, abundant food sources, a variety of pools and riffles, plenty of places to hide and clean, cool water are all important elements of a healthy river habitat.

The riparian zone includes the trees, shrubs, grasses and groundcovers that fringe the river and influence—and are influenced by—water. Research has shown that healthy riparian zones are occupied by more species of fish and wildlife than any other type of land. Standards for the protection of riparian zones have been adopted in Washington and other states.

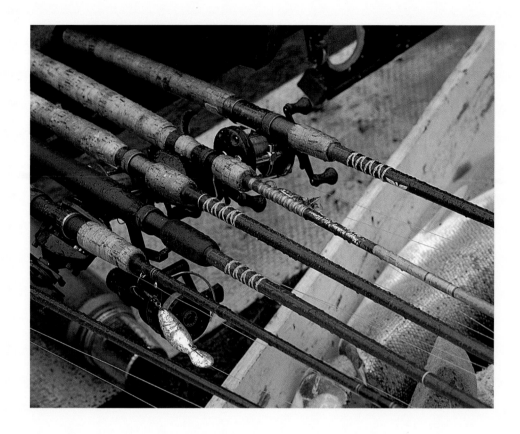

Indicator species are certain animals and plants whose presence or absence can be used to
    determine the health of a watershed.

River banks lie between the water's edge and adjacent ground. Here, roots bind the soil together
    and plants protect against erosion. A *river bed* is its bottom, usually covered
    by a mixture of sand, gravel, silt and smooth stones.

The portion of a river bed in which the water is flowing is called the channel. The water that
    people withdraw from a *channel* is known as the *offstream flow.* The water that remains
    after such withdrawals is the *instream flow.*

Pools are deep, scoured portions of a river or stream, where water flows slowly. These can be
    feeding and resting places for salmon and other fish.

Riffles are shallow rapids where water flows swiftly over rocks and gravel, adding oxygen to the
    water and stirring up any food particles on the bottom.

◄ AND ▲ *Rocks and rods: familiar*
elements of watershed life.

35

ALCHEMY

ALCHEMY

Origins of the Upper Watershed

For the first droplets of the Nisqually River water, there is nowhere to go but downhill. The thin topsoil cannot absorb the tiny pearls of moisture, and the sparse plants lack amply developed root systems to draw them in. Thimble-sized drops coalesce, trickling in rivulets down the slopes of Mount Rainier. Rivulets merge, forming small streams.

Each new stream carries tiny particles of rock, the pulverized aftermath of the glaciers' ebb and flow. No bigger than plaster dust, these particles are called "flour"—a fitting name for any material produced by the mountain's grindstone and the weight of the Nisqually Glacier's slow-moving mill wheel.

Tinged the color of chocolate milk, the flour-laden water sweeps over a field of cobbled rocks deposited in previous eons at the glacier's foot. The young river barely keeps to its cobbled bed in this steep upper stretch. It jostles and tugs at each rock in its path, coursing ahead with the youthful exuberance of a new river.

The energetic stream finds other new snowmelt sources, some milky and some unusually clear due to non-glacial origins. As it tumbles downhill, the stream divides into two parallel streams for a stretch, then soon reunites in a pattern known as "braiding" for its strong resemblance to plaited hair.

◄ ◄ (PREVIOUS PAGE)

Narada Falls,

Paradise River.

◄ *A bridge on the aptly*

named Wonderland Trail

carries hikers across the

Nisqually River in Mount

Rainier National Park.

GAINING STRENGTH AND SPEED

The young Nisqually joins other swift flows near the entrance to Mount Rainier National Park. The merging waters carry a cargo of leaves, needles, bits of bark and logs from the thick evergreen forests lining the river banks. The pillars of this old-growth forest community are tall, thick-barked Douglas fir, western hemlock and red cedar, some more than 700 years old. While these stately trees must endure the mountain's long, severe winters, they are no longer susceptible to the woodcutter's axe—a reprieve granted in 1899 with the creation of Mount Rainier National Park.

By establishing the park's boundaries, which today encompass a total of 235,612 acres, the federal government protected much more than trees. Park regulations limit development and restrict the range of visitor activities to non-consumptive outdoor recreation and the passive study of nature, thus minimizing human disturbance and most water quality threats to the river's source and flow. To safeguard other water sources, a range of protective measures have also been put in place at the headwaters of many rivers and streams in our nation's more densely populated watersheds.

Emboldened by the influx of waters from rivers and streams and the steeply sloping terrain, the Nisqually surges powerfully ahead. In shallow stretches, especially those full of bends, the friction of water against rock and riverbank momentarily slows the river. In straight channels, where waters run deep and narrow, the river gathers speed. Fast-moving water quickly erodes the banks and gradually softens the rough edges of boulders and other obstacles. It also carries greater amounts of larger drift material—rocks, sand and silt—than do

▶ Within Mount Rainier National Park, a grove of ancient evergreens provides shade along the road to Longmire.

slower-moving waters. Whenever the rushing river loses momentum, the drift material drops out. Shoved aside by the current, it mounds into islands, shoals and bars.

Some Watershed residents, such as the water ouzel or dipper, capitalize on the niches made by eroding banks and shifting sandbars. Showing no fear of swift water, this robin-sized bird plucks caddisfly larvae (a seasonal delicacy) from among the cobbled rocks on the river bottom. To nab such tidbits, the dipper plunges into the water, spreads its wings and "flies" in the fast-moving current. Having captured its prey, it pops out of the current and onto a boulder or streambank to eat.

Dippers build nests made of mosses beneath bridges and overhangs. Like many of the animals that feed or breed in Northwest watersheds, this intrepid avian favors healthy riverside habitats and reliable sources of clean, fast-moving water.

Other Watershed dwellers have benefited from the Nisqually River's rough temperament and wandering ways, including the Squalli-absch people, ancestors of the modern Nisqually Indian Tribe. The Squalli-absch followed the river's rough-hewn path through the oftentimes impenetrable wilderness, thousands of years ago. According to legend, they came north from the Great Basin, crossed the Cascade Mountain Range and erected their first village in a basin now known as Skate Creek, just outside the Nisqually River Watershed's southern boundary.

Centuries later, bands of the Squalli-absch built villages throughout the Watershed, moving progressively farther downriver—from what is now the town of Elbe to a site near the Mashel River. Then they ventured out of the forest and onto the prairies near present-day Roy and Yelm. The members of these

◄ *Near Elbe, the persistent river sculpts rocks and moves trees.*

► (FOLLOWING PAGE) *The Nisqually Indian Isalahah (also known as Indian Sam) plied the river with his cedar canoe. In 1890, he was granted a homestead patent for 160 acres along the Nisqually River.* (BACKGROUND) *Reflection Lake.*

43

bands gave the river its name—Squalli, their word for the tall grasses that covered the plain.

Working the prairies, woodlands and rivers, the first people of the Nisqually harvested berries and herbs, hunted ducks, grouse, deer and elk, caught fish and collected freshwater mussels and clams. Every activity in their villages was guided by the concept of reciprocity—the give-and-take between humankind and the rest of the natural world. All animals and plants had spirits, setting them apart from mere commodities. The majestic red cedar tree, from which the tribe collected raw materials for baskets, clothing, canoes and homes, was especially sacred. Equally venerated was the salmon; villagers would gather to give thanks and to celebrate the yearly return of this important food and esteemed friend. Similar ceremonies honored land and sea mammals, such as elk and seals.

Nisqually Indian baskets are functional and pleasing to the eye, with woven patterns that symbolize mountains, salmon, people and other Watershed features.

FOOT TRAILS AND IRON RAILS

Subsequent non-Native travelers arrived in the Watershed by boat, moving inland by foot from the mouth of the Nisqually and following the same water-carved trail. They, too, built villages throughout the Watershed, never far from the rushing water that granted them access to the interior. They quickly discovered many of the same resources used by the Squalli-absch and, in many instances, learned to respect the wild spirit of the land.

Several leaders of this second wave of settlers were led by a Yakama Indian named So-to-lic, known to European pioneers as Indian Henry. It was So-to-lic who, in the early 1880s, befriended James Longmire, founder of the Longmire Springs Hotel and Baths. With the establishment of Mount Rainier National

▶ A vacationer enjoys the calm between winter snowstorms at Longmire.

46

Park, the hotel site (now bearing the name Longmire) became a hub for visitors to the Nisqually River Watershed. Between 1899 and 1903, about 2,000 people ventured to this part of the Park; by 1906, nearly that many came in a single season.

The availability of fish, timber, minerals and other resources and the ease of removing them to neighboring communities with larger populations directed the settlement of the Nisqually and other Northwest watersheds. The rapid development of the Ashford/National area, some six to eight miles west of the Nisqually entrance to Mount Rainier National Park, typifies the "boom and bust" cycles that drove many small upper Watershed towns. This community

In operation from 1906 until 1926, the original inn at Longmire hosted thousands of dedicated travelers. Its successor, the National Park Inn, continues to serve visitors to Mount Rainier.

❧ ▲ *Removed from its original site at National, a restored bunkhouse now greets people in Ashford.*

came to life in the late 1800s, when homesteaders staked their claims, cleared brush and timber and began tilling the land. Other bold entrepreneurs soon moved in to capitalize on the largely untapped resources of the richly forested region. One of the most significant arrivals was the Tacoma Eastern Railway, a subsidiary of the Chicago, Milwaukee and St. Paul line, which brought new opportunity to the wilderness settlers in 1904.

With communities linked to new markets by the railroad, subsistence farming was soon supplanted by large-scale timber harvesting and log shipping. Sensing the change that rail traffic would bring, businessman Walter Ashford platted his town on August 7, 1904. It became the terminus of the railroad, complete with a hotel that could accommodate 20 travelers.

The following year, National was established as a company town for the Pacific National Lumber Company. With its massive sawmill and peak population of 4,000, the town could be seen from neighboring Ashford, roughly two miles away. Even though fire consumed the mill and much of the town in 1912, it failed to slow the pace of production. The mill was soon rebuilt and work continued at a brisk clip. The flavor of the town in its heyday was neatly recorded by the federal Works Progress Administration in the 1930s:

"The great red buildings of the mill, its rusted stacks belching black smoke and white steam, dominate the town. Crowded close together and fronting cracked, planked streets are tiny box-like cottages painted in the same dingy red as the mill. The lumber company dominates every phase of the town's activity, and no one who does not gain his living through the business of the mill lived in National."

◄ ▲ *Locomotive Number 11 was one of several steel workhorses in the service of the St. Paul & Tacoma Lumber Company in 1929.*

◄ *An automobile and a passenger train raced from Tacoma to Ashford in 1910, ending in victory for rubber tires.*

▲ *Today, sight-seeing trains of the Mount Rainier Scenic Railroad chug along tracks from Elbe to Mineral and back.*

49

These prosperous times were short-lived. In 1944, the sawmill at National was acquired by the Harbor Plywood Corporation, which, in turn, sold its holdings to Weyerhaueser Timber Company in 1957. By then, neither National nor Ashford was a logging center, and Weyerhaueser soon traded these holdings to the Washington Department of Natural Resources for other Watershed lands. The same railway that once carried timber down the mountainside now ferried tourists to and from Mount Rainier National Park.

Travel to the upper Watershed was extremely difficult for most pioneers—holding back would-be prospectors, farmers and loggers. The original wagon road from the outskirts of Tacoma to Longmire Springs was so steep and deeply

◄ ◄ Eroding fragments are all that remain of National's timber heritage.

◄ A moment of calm at the Pacific National Lumber Company is captured in this rare picture from the 1940s.

▲ A pair of modern conveniences, blacktop and big trucks make it easier to haul timber in the Watershed.

rutted in places that at least three good horses were needed to pull a wagon uphill. Whenever feasible, early settlers layered logs crosswise on roads, creating corduroy roads that badly jostled spines but provided better traction than unimproved surfaces.

With interest in Mount Rainier National Park on a steady rise, federal and state officials made a concerted effort, beginning in 1903, to smooth the way to the upper Watershed. It was not an easy task. Hampered by the region's glacial heritage and the many floods, rockslides and mudflows that continually reshaped the terrain, work crews took more than a decade to complete the highway. The fruit of their labor—a narrow but fully paved roadway from Tacoma all the way to the campgrounds at Paradise—proved invaluable to motorized movement within the Watershed.

◄ ◄ AND ◄ *The road to Paradise, now and then.* ▲ *"A great many had grown weary of the way so steep and dreary," observed the* Elbe Union *in 1896. Road improvements between Tacoma and Paradise were soon to follow.*

A Creeping Giant

THE NISQUALLY GLACIER is the seventh largest of the 25 major glaciers on Mount Rainier's broad shoulders, and one of six flowing down from the summit. As with large ice masses throughout the world, its shape changes over time. It grows larger and moves downslope during cooler decades with heavy snowfall. It shrinks and retreats uphill in warmer decades with light snowfall. Advances and retreats permanently engrave the landscape with deeply plowed canyons and litter it with rocky debris. Near the upper surface of the Nisqually Glacier lies a layer of thicker, heavier ice created by many years of deeper than usual snowfall. This ice actually flows downslope like a slow-motion wave, eventually pushing the glacier's tip forward when it reaches the terminus.

During the Pleistocene Ice Age, the Nisqually Glacier pushed its rocky debris beyond the current town of Ashford, extending itself as a single sheet of ice, in some places 1,600 feet thick. In warmer times, the glacier has retreated, moving back up the mountain as much as 10 inches in a day. These slow but steady movements give glaciers their nickname—rivers of ice.

By studying the rocks and reading the records of early explorers and pioneers, we've learned that the Nisqually Glacier has moved backwards and forwards many times over its long life. In his journals from 1857, the mountain-climbing

Lieutenant August Kautz of the U.S. Army wrote that the glacier's tip reached the location now occupied by the highway bridge leading to Paradise. Rock debris left by the glacier tells us that, centuries before, the tip extended 900 feet below the bridge. In the 1890s, homesteader James Longmire complained that his trips to cut household ice were getting longer each year—clearly the Nisqually Glacier was receding in those days as well. In the 1920s, visitors to Mount Rainier posed for photos using the glacier as a backdrop. Thirty years later, this photo point was still popular, but the Nisqually Glacier had moved out of the picture. Now it was more than a mile up from the highway bridge.

Concerned that the Nisqually Glacier was slowly disappearing (and taking with it snowmelt that runs hydropower plants downstream), the Tacoma Light Department and the U.S. Geological Survey in the 1930s began taking annual measurements of the glacier's ice. Their measurements confirmed that it was normal for this river of ice to ebb and flow in response to the ever-changing climate of the Pacific Northwest.

◄ ◄ *Nisqually Glacier.*

◄ *Photo opportunity, circa 1920.*

Where railroad and roadway intertwined, the town of Elbe was built, one of the oldest non-Indian settlements within the Nisqually River Watershed. German immigrants named the town after the Elbe River, a beloved feature of the land they had left behind. Levi Engel, community blacksmith and editor of the *Elbe Union* newspaper, ran the first garage and service station in the area, selling gasoline from a barrel. Owing to the small capacities of automobile gas tanks in those days, nearly every car en route to Mount Rainier made a stop at his Elbe fuel pump. Engel's first customer was also his most famous—President William Howard Taft, who was taking a trip to Mount Rainier National Park in October 1911.

HARNESSING THE RIVER

Reliable roads and rail service suddenly brought the resources of the upper Watershed within reach. Large-scale engineering projects, once impossible to implement, could now proceed. The largest of these involved the original La Grande Dam, a hydroelectric project started in February 1910, that took only two years and a major commitment of $2.4 million in cash to complete.

Constructed by what was then a rapidly growing Tacoma Light Department, the finished dam temporarily impounded the Nisqually River behind a 45-foot wall of concrete. Around 900 cubic feet of water per second was diverted from the Nisqually, routed through a 1,300-foot-long settling channel, then into a tunnel carved into solid rock more than two miles long. From there, it was carried by four large penstocks to a staunchly built powerhouse below. Within the powerhouse, four sets of Allis-Chalmers turbines and generators

▶ Under construction in 1942, Alder Dam dwarfed its predecessor, La Grande Dam. Completed in 1944, this new facility provided electricity to heat and light 21,000 Tacoma homes. (BACKGROUND) *La Grande Reservoir.*

turned fluid movement into electrical energy. Transformers "stepped up" the power from this machinery to 60,000 volts, which was transmitted to the Tacoma Light Department's Nisqually substation.

"Everything was rosy" during its first year of operation, reported Dick Malloy in *The Tacoma Public Utilities Story: The First 100 Years.* "The city's demand was 20.4 million kilowatt-hours per year and the Nisqually hydro project was supplying it all." With hundreds of new customers clamoring for the latest electric conveniences, more attention was focused on the Nisqually River Watershed—and its potential for hydropower.

The demand for electricity from the Nisqually River became even stronger in the years before, during and immediately after World War II. For Tacoma's leaders, there was only one answer: build more dams. With a 50-year license from the Federal Power Commission, Tacoma City Light embarked on an even more ambitious project on the Nisqually, constructing a larger facility, Alder Dam, downstream from the original La Grande Dam. Generators within the new dam's 330-foot arch could produce 50,000 kilowatts—enough electrical energy to heat and light 21,000 homes. An additional dam (also called La Grande) was constructed downriver to restrain any water leaving Alder Dam for a second time. Water would then pass through a tunnel to generate another 64,000 kilowatts at an enlarged powerhouse. Coupled, the two improvements would increase City Light's capacity by 80 percent.

Completed in 1944, the Alder-La Grande hydropower complex initially provided Tacoma residents with "peak power"—extra energy for times of high electrical use. Providing peak power meant regularly releasing additional water for generating electricity when demand was highest, usually in the early morning

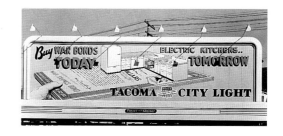

◀ *Concrete arch, Alder Dam.*

▲ *The war effort brought a flood of new workers to Tacoma. "In turn, utility department employees are being kept busy as woodpeckers on city light poles, making service changes at the rate of 50 per day," reported the* Tacoma Daily Tribune.

and evening. The practice, which continued until 1968, wrought dramatic changes to the river and its adjoining lands. In some stretches, the channel was severely scoured, in others it lay nearly buried by clay and gravel eroded from its banks. Once-healthy riverbank habitats were routinely immersed and then left to dry, with disastrous consequences to all but the most adaptive animals and plants.

When another hydroelectric project was completed by the City of Centralia Light Department in 1930, it, too, influenced parts of the Watershed as far as 14 miles downriver. Periodic water releases from the Centralia project have abraded the channel of Thompson Creek and piled up gravel and rock, creating a large delta that today reaches halfway across the Nisqually River. A monument to human impact on the environment, the delta has become a popular spot for picnicking, wildlife viewing and fishing.

The Alder-La Grande hydropower complex added a significant feature to the Watershed—Alder Lake, the seven-mile-long reservoir for Alder Dam—while at the same time taking one away. Resting well below the reservoir's surface are the original La Grande Dam, 250 acres of former farmland and the last remains of Alder, a town named for the thick stands of trees that encircled the site. Once a busy mill town and agricultural community, Alder prospered through the late 1920s, until the end of rail service caused business to decline. In 1941, when Tacoma City Light announced its plans to purchase properties for its new hydropower project, most residents jumped at the opportunity to unload their surplus lands. Still others rejoiced over the jobs the project would bring. Most of Alder's houses were dismantled before work on the dam was completed. The schoolhouse and church were moved uphill to a new townsite, which today bears the Alder name.

▶ Low snowmelt makes a mosaic on the fringe of Alder Lake.

THE HIDDEN COST OF HYDRO

All three hydropower projects benefited communities beyond the Watershed. Yet because of the interconnected dynamics of Watershed life, they unintentionally harmed the river and its drainages in several ways. Particularly hard hit were the five native species of salmon—the pink, coho, steelhead, chum and chinook.

In its first years of operation, Centralia's diversion dam lacked a fish ladder to facilitate upstream or downstream migrations of these commercially important fishes. A number of salmon runs, especially pink and chinook, were stopped by the dam during the driest weeks of summer and early fall, when barely enough water for the salmon's upstream journey spilled over the dam. And from 1930 to 1955, the dam had no fish screen on its diversion canal. Many sea-going offspring of the salmon that successfully cleared this obstruction were dispatched by the whirring blades of powerhouse turbines.

Attempting to revitalize the rapidly dwindling chinook and coho salmon runs, the Washington Department of Fisheries began releasing juvenile fish in the river during the 1940s. Around the same time, the Washington Department of Game augmented the river's native steelhead runs by introducing a non-native strain—the so-called winter steelhead, which now instinctively returns to the Watershed from November through April. The winter steelhead migration is the largest run of its kind in the entire Puget Sound area.

In 1955, Centralia renovated its hydroelectric facility, raising the height of the diversion dam and nearly doubling its generation capabilities. In a mitigative move, Centralia also added fish ladders and protective screens. In 1975, the Nisqually Indian Tribe sued Centralia, seeking further restitution for the runs damaged by the diversion dam operation. A similar petition a year later sought to undo the harmful effects of Tacoma's Alder-La Grande operation. From these proceedings came mutually acceptable standards for determining how much water could be redirected for hydropower generation during the different seasons. Additional agreements were made with the cities of Tacoma and Centralia to help support the Tribe's salmon hatchery program.

Several fish populations responded dramatically, once the standards were adopted and the river's flow was stabilized. Pink salmon runs, which had dropped to fewer than 200 returning fish, burgeoned to a peak population of 20,000. But these improvements came too late for a strain of native chinook that spawned each spring in the upper reaches of the Watershed and was especially susceptible to low flows. Historic records suggest that the last spring chinook in the Watershed vanished by the late 1940s.

In 1977, when the state Department of Fisheries built the McAllister Creek hatchery to raise chinook salmon, primarily for the Puget Sound sports fishery, the Nisqually Indian Tribe constructed a facility of its own, the Kalama Creek hatchery, designed to bolster sagging coho and chinook runs. The Tribe also began a program to "plant" fingerlings of chum salmon in Yelm Creek and, in years of low flow, Muck Creek. The Fort Lewis Military Reservation, one of the major landholders in the Watershed, provided land for a second hatchery at Clear Creek. Congressional funds for construction and start-up were administered by the U.S. Fish & Wildlife Service, and the City of Tacoma paid for operation and maintenance. Dedicated in 1991, the hatchery serves as a model of inter-governmental cooperation—a key ingredient in restoring the Nisqually River Watershed's salmon runs.

Since the mid-1970s, the Nisqually Indian Tribe has taken the lead in cooperative efforts to manage the Watershed's salmon stocks.

CONFLUENCE

Meetings in the Middle Watershed

Below La Grande Dam, the Nisqually River enters a narrow canyon, its walls of volcanic rock held in place by densely packed gravel and hardened clay. Rimming the canyon are dark stands of shaggy Douglas fir, red cedar and red alder. The bobbled reflections of tree limbs in the brownish green water, the arching fronds of horsetail at the water's edge—both signal that the river, while silent, is still on the move.

For most of the year, this stretch of the river is neither milky with glacial flour nor tinged with decaying plant matter from the forest. Below the City of Tacoma dams, soil particles and microscopic green algae color the waters, well stirred by a succession of stairsteps, rapids and riffles. Here, the river is wider than ever, nearly 60 feet across. Its channel curves in a sinuous path, first arcing northwest, then northeast as it winds toward a second deep chasm—the canyon of birds.

The north face of this second gorge is pockmarked with nest holes of rough-winged swallows—small, brown-backed, white-breasted migrants that overwinter in South America and return to the Nisqually each spring. Cliff swallows also build their nests here, fashioning gourd-like jugs of mud that adhere to the almost vertical surfaces. Both species of long-distance fliers gather throughout the middle Watershed to nest in colonial roosts. Within this canyon alone, more than a hundred pairs have been counted in one year.

◄ ◄ (PREVIOUS PAGE)

Reflected light on Ohop Creek.

◄ *No longer in use, a narrow loggers' bridge spans Ohop Creek.*

Nestbuilding for rough-winged swallows begins in April. Birds occupy holes made in previous seasons or dig new ones in the canyon wall. Lining their nests with grasses and feathers, they're soon tending four or five small white eggs. After the young leave the nest, a pair may have a second brood. Youngsters from the first broods often form flocks, perching together and singing in the same chittering voice as their parents.

MASSACRE ON THE MASHEL

Leaving the bird cliffs behind, the river passes beneath the remains of a small bridge. Once a major crossing for trucks bearing freshly cut logs, it forms a decomposing substrate for an assortment of algae and lichens. Beyond the bridge, the river widens and slows even more, as if in anticipation of the next juncture, the confluence of the Nisqually and the Mashel rivers—a meeting of the waters.

This meeting place has always held special appeal to Watershed dwellers. Not far from here, thousands of years ago, the people of the Nisqually Indian Tribe built Me-schal, the largest of three permanent villages in the shadow of Mount Rainier. Me-schal villagers were intermediaries, living links in a long chain of trade ties and blood relations that connected tribes on both sides of the Cascade Mountain Range. Their settlement continued to serve as a vital Indian center until the Puget Sound Indian Wars of 1855-1856.

A series of tragic events led to the village's doom. Trouble began in December of 1854, when Isaac Stevens, first governor of the Washington

Both sexes of cliff swallows contribute to the construction of mud nests and line them with downy feathers—perhaps from their own breasts.

Territory, negotiated a treaty with the Nisqually, Puyallup and several smaller Indian tribes. Under this treaty, the Nisqually Indian Tribe was expected to surrender all of its holdings except a small wooded hillside with particularly thin and unproductive soils. These terms were unsuitable to the brothers Leschi and Quaymuth, residents of Me-schal village and leaders of the Nisqually Indians. When an order to take the brothers into custody was issued, the tribe vowed to fight back. Several skirmishes erupted between the governor's troops and the Indian warriors, a party of about 150 men.

The stakes were upped in January 1856, when Stevens issued an order to exterminate all hostile Indians. That April, Captain Hamilton J.G. Maxon of the Washington Mounted Rifles and his battalion of 58 men captured an Indian on the Ohop drainage, forcing him to lead them to the others hiding at Me-schal. Maxon launched a surprise attack, killing a large number of old men, women and children.

Eventually Governor Stevens negotiated a peace with the "friendly" Nisqually and Puyallup Indians, held captive on Fox Island, near Tacoma. Leschi was taken into custody, tried as a civilian, not a soldier, and convicted and hanged on February 19, 1858. Quaymuth surrendered himself to the authorities and was murdered as he slept on the floor of the governor's office in Olympia. His murderer was never apprehended.

After the Puget Sound Indian Wars, many tribal members were removed from ancestral home sites and taken to reservation lands. But not every Nisqually went along with the plan. "A great many Indian people got lost in the shuffle," concluded Cecelia Svinth Carpenter in *Fort Nisqually: A Documented History of Indian and British Interaction*. Many Nisqually families may have chosen

◄ (PREVIOUS PAGE) *Fed by the Little Mashel, a tributary that collects water from 25 square miles, the Mashel River supplies much of the Nisqually River's instream flows.*

▲ *A posthumous portrait of Chief Leschi.*

► *Nisqually Indians called their mountain Ta-co-bet, or "nourishing breasts," after the life-giving waters that flowed down from its slopes.*

to reside on the nearby Puyallup Reservation. "Other Indian families roamed the woods, found a place to squat and lived out their lives outside of the reservation," wrote Carpenter. Some Nisqually Indian women married white pioneers, moving to other parts of the Watershed and adopting new roles as wives of farmers and herdsmen.

The Nisqually Indian Reservation was considerably larger than that originally rejected by Chief Leschi. Spread over approximately 5,000 acres, this new property included river frontage for fishing and prairie land for grazing horses. The boundaries were drastically reduced in 1917, when the Pierce County government condemned nearly two-thirds of the Nisqually Reservation, annexing it to nearby Fort Lewis. Today, reservation land encompasses only 1,650 acres in the middle Watershed. Housing projects assure homes for a large segment of the tribe, and the Nisqually Indian center includes a library, senior citizen facility, law enforcement building and an active natural resource division to protect and improve salmon resources.

Red-legged frogs are highly susceptible to changes in the environment, making them good indicators of watershed health.

FORESTRY'S FRESH FACE

In 1889, not far from the confluence of the Nisqually and Mashel, So-to-lic met land speculator Thomas C. Van Eaton and convinced the newcomer to put down roots on the Mashel's riverbanks. Van Eaton purchased 100 acres, 60 of which he eventually gave away. He opened a general store, outfitting homesteaders in the nearby Mashel and Ohop valleys and, in later years, adding a post office to the building that housed his store. The area's first school was also

built on a parcel of land donated by Van Eaton. The general store and log schoolhouse, a cafe and a few other structures attributed to the town's founder are still standing—some rather unceremoniously disguised by modern plywood siding and asbestos shingles—in Eatonville's central business district.

Like the timber towns of Ashford and National, Eatonville grew rapidly, then gradually declined. Its population soared in 1907, when a group of midwestern investors established the Eatonville Lumber Company. The Tacoma Eastern Railway also chose Eatonville as a terminus for its line serving the many small logging communities in this part of the Watershed. Suddenly, instead of 70 permanent residents, more than 300 called Eatonville home. Most worked for the lumber company, which, at its peak, stacked about 10 million board feet of wood in its yard.

Accepting its twin roles as mill town and agricultural supply center, the town prospered and grew. Only when harvestable timber became scarce in the mid-1950s did production grind to a halt, shutting down the lumber company. Despite the mill's closure, Eatonville businesses still do a healthy business supplying area farmers, builders and logging concerns.

Logging continues to shape the identity of the middle Watershed, although the actual practice of harvesting trees has changed substantially over the years. Nearly two-thirds of the Watershed's acreage is timberland, owned and managed by a range of interests—including the U.S. Forest Service, state Department of Natural Resources, University of Washington and, in the private sector, Weyerhaeuser, Murray Pacific, Champion International and Plum Creek. The Nisqually Indian Tribe and U.S. Army also have holdings of forested lands, much of which is managed for timber production.

◄ ◄ (PREVIOUS PAGE) *A buffer of trees encircles Tule Lake, nestled between Harts Lake and Tanwax Creek.*

◄ *Bison and other domestic beasts enjoy Eatonville's clean water and ample pasturage.*

75

A mural on the west wall of the local Chamber of Commerce celebrates the glory days of Eatonville's lumber mill.

Turn-of-the-century loggers take a break from hard labor to pose for a photo. (BACKGROUND) *A freshly cut log.*

The first loggers in the Nisqually River Watershed could afford to remove the biggest, straightest and best timber and leave the rest. After taking the most desirable resources in one part of the Watershed, work crews simply moved on to the next. Little thought was given to replanting these harvest sites or stabilizing the freshly exposed forest floor. These early operations unintentionally invited wind and rain to erode the top layers of soil, which often washed from the land into the Nisqually's feeder streams. Here it smothered salmon eggs and damaged fish and wildlife habitats. Among the hardest hit was the Mashel River, whose narrow channel leaves it vulnerable to both human impacts and natural occurrences. Historically, the Mashel's uppermost reaches have been alternately scoured by flash flooding and buried by silt and woody debris from logging operations.

A significant change took place near the turn of the century. Understanding that timber supplies in the Watershed were finite, some foresters began to take steps to ensure a second harvest in the future. They retained title to the land and took steps to repair and replant the forest.

One of the largest private timberland owners in the Watershed, the Weyerhaeuser Company has also been the most active supporter of the shift toward sustainable forestry. In 1900, the timber giant began purchasing land for the Vail Tree Farm, an ample portion of which extends into the Mashel River drainage. Steadily adding to its holdings at Vail over several decades, Weyerhaeuser started to operate the first large-scale timber plantation in the Watershed in 1946.

Over the past two decades, federal and state regulations and private work standards have greatly improved operations along the Nisqually and its tributaries. Research conducted at facilities like the Charles Lathrop Pack Experimental Forest (described on page 81) has significantly advanced resource management techniques and made sustainable harvesting of timber an achievable goal throughout the Nisqually River Watershed.

Tree seedlings are routinely hand-planted in freshly logged tracts—a proven technique for helping the forest regenerate. At the Vail Tree Farm and other harvest locales, streamside timber stands are now left uncut to enhance the fisheries in these waters. Fallen logs and root masses are also left in place, attracting woodpeckers, chickadees and other wild birds that excavate cavities and use them for nesting. Other birds and small mammals seek insects that live under the bark and in the wood. As the wood decays, mosses, liverworts, lichens and fungi grow and flourish—important food sources for some of the forest's largest and tiniest animals.

◄ *A forest of mostly 60-year-old trees covers the land near the town of McKenna.*

▲ *For many decades, forestry students have acquired valuable hands-on skills at Pack Forest.*

Protected Parcels of Diversity

WHILE THE HEALTH of the upper Watershed's waters is attributed to protection conferred by Mount Rainier National Park, the middle stretch of the Nisqually River owes much of its well-being to safeguards that stem from several smaller public holdings. One of these is Northwest Trek Wildlife Park, an unusual oasis north of Eatonville. Originally the property of the Northern Pacific Railroad Company, this land was sold to the St. Paul & Tacoma Lumber Company (St. Regis) in 1914, changing hands once again after 20 years of logging and a disastrous forest fire.

The tract's new owners, Dr. David and Connie Hellyer, raised cattle and ran a tree farm from 1958 until 1971—when they got the idea to turn their property into a wildlife preserve. The Hellyers deeded their land to the Metropolitan Park District of Tacoma, with the stipulation that

David be given "a planning and supervisory position and a nominal salary (one dollar per year) to assist in the development of the park." Released from the pressures of production, the land rapidly rebounded. Northwest Trek opened to the public in July 1975, providing visitors with exhibits of Northwest wildlife, walking trails, naturalist-guided tram tours and 435 acres of second-growth forest and meadow turned over to free-roaming hoof stock such as elk, bison, moose and mountain goat.

An equally positive influence in the middle Watershed began in the 1920s, with financial assistance and philosophical guidance from the enlightened East Coast lumberman, Charles Lathrop Pack. With Pack's support, the University of Washington developed its experimental forest—a place for researchers to improve forest management techniques, for university forestry students to acquire practical field experience and for timber owners, loggers and the public to observe examples of sound forestry practices year-round.

Early research at Pack Forest focused on plantings, with college students helping to install plots of western red cedar, Douglas fir, ponderosa pine and Port Orford cedar. After World War II, Douglas fir seedlings from Pack Forest were shipped to Japan, assisting in the rebuilding of war-scarred watersheds overseas. The native and exotic trees at Pack Forest, some now nearly 70 years old, are the subjects of continued research. Currently spread across more than 4,000 acres, the experimental forest contains representatives of many native and non-native tree species. Their contributions toward the healthful diversity of wildlife habitats within the Nisqually River Watershed should not be overlooked.

A third haven in the middle watershed, Bald Hill Natural Area Preserve (NAP) covers approximately 290 acres and encompasses Bald Hill Lake and a canyon east of the lake. However, this site's principal attributes—an old-growth forest, Oregon white oak woodland and populations of four plants listed as "sensitive" in Washington State—could easily suffer damage from a steady stream of visitors. In 1985 and again in 1987, The Nature Conservancy and Weyerhaeuser entered into agreements to set aside an area of old-growth timber, wetlands and grasslands for limited research and education. The NAP is now managed by the Natural Heritage Program of the Washington Department of Natural Resources.

Additional Watershed acreage has been set aside by the Nisqually River Basin Land Trust, a nonprofit group that protects land through acquisition. Working from a "wish list" of sites, the land trust and its partners in the public sector are methodically obtaining large undeveloped tracts of Watershed to preserve as habitats for wildlife and to provide recreational and aesthetic benefits for people.

◄ ◄ *A shaggy resident of Northwest Trek.*

◄ AND ▲ *The experimental forest and its namesake, Charles Lathrop Pack.*

81

ENTER THE OHOP

Two miles south of the Mashel, Ohop Creek meets the Nisqually River in another important confluence. Here, silt and soluble nitrogen, flushed by rainwater and snowmelt from pastures and farm fields upstream in the Ohop Valley, add new colors and tastes to the olive-hued river water.

These additives probably first appeared in the Ohop shortly after 1874, the year Robert Fiander, the region's first homesteader, filed a land claim to the Ohop area. Other settlers followed Fiander's lead, including Norwegian farmers like Torger Peterson and his family, who had survived a month-long crossing of the Atlantic Ocean and a train ride of several weeks to arrive at this particular place in the Watershed. Peterson, along with his friends Ole Halverson and Herman Anderson, performed the first land survey of the Ohop Valley, selecting three choice parcels of potential farmland for themselves.

As other homesteaders moved into the valley, they converted the forests to farmland and planted a full spectrum of crops. The business of providing food for the loggers and their draft animals proved lucrative for some. Many bartered for items they couldn't grow themselves—including oats for their livestock, obtained from So-to-lic and his three wives. Bartering helped bring Europeans and Nisqually Indian Tribespeople closer together, erasing some of the emotional scars collected during the Puget Sound Indian Wars.

"One of the highlights that comes to my mind is the first Indian Salmon Bake in the community," recalled long-time Ohop Valley resident Matteus H. Kjelstad, in a warmly narrated attachment to *Highway to Paradise* by Gene Allen Nadeau. "The Indians invited the settlers in the valley to this affair set on the

The original rails of the St. Paul & Tacoma Lumber Company lead downvalley, past a vivid stand of vine maples.

banks of the Nisqually River about a quarter mile upstream from the mouth of Ohop Creek, where there was a fine beach with overhanging alder trees."

The spring run of chinook salmon had just begun, according to Kjelstad, so on the morning of the get-together, three Indian fishermen launched their canoes, traveling upriver with a hundred-foot net. "They string the net across the river in places where there were no obstructions," Kjelstad remembered. "It was then unloaded of fish and the same process repeated at the next open space down the river. When they arrived at the picnic site, they had about 15 fine shiny chinook salmon in the canoe." The salmon feast became a yearly event on the Mashel Prairie and often included a baseball game between Native and non-Native players.

Sometimes Ohop Creek overran its banks, muddying the lines between agriculture and aquaculture. Quick to capitalize on any opportunity, Ohop farmers gaffed the salmon that lay stranded in their potato fields and, in the words of local historian Helen Danforth, "brought back a few spuds as well for a fish and chips dinner."

◄ AND ▲ *A place of living history, the Pioneer Farm Museum offers insights into the lives of the Ohop's first farmers.*

A FARM LEGACY

While most of the Ohop's original farm families left the valley long ago, their legacy lives on at the Pioneer Farm Museum, two miles west of Eatonville. Among the intriguing features of this interpretive center are a pair of homesteaders' cabins and a trading post. Built in the 1880s and weathering many winters in the Watershed, all three were moved to Pioneer Farm and restored

Settlers of Harts Lake cleared the land, tilled the soil and constructed a one-room school for their children to attend.

The legacy of family farming lives at Harts Lake.

by the Museum's staff. Visitors to Pioneer Farm are given chances to churn butter, pump the bellows of a turn-of-the-century blacksmith's forge and play in a hayloft—all excellent opportunities to experience a little of the lives of the Watershed's early pioneers.

Agriculture practices still color the Ohop Valley and other niches in the Watershed. However, for the most part, the focus has shifted away from raising many different crops and a few head of cattle on the same farm. Now the majority of agricultural lands in the middle and lower Watershed are devoted to livestock and dairy production.

As in many parts of the Northwest, the complexion of farming in the Nisqually River Watershed is changing in other ways. Many large commercial

farms are being converted to small, non-commercial operations, commonly known as "hobby" or part-time farms. The combined effects of these small farms on water quality may be greater than those of well-managed, commercial operations, typically 25 acres or larger. Part-time farmers often lack the experience, training and space to properly manage their lands or put pollution controls in place. A 1994 inventory of farms in the Mashel and Ohop Creek drainages found half of these operations on less than 10 acres. Ironically, the highest animal densities were found on the smallest farm sites—a common situation in many watersheds with agriculture in transition.

Overcrowding puts undue pressure on pastures and stock and also compromises water quality. Washed into rivers and streams, nutrients from animal waste can lead to an imbalance in the natural nutrient cycle, robbing fish and other aquatic life of the oxygen they need to survive. Animal wastes are also sources of pathogens—bacteria and viruses harmful to human health. Along with pesticides and petroleum products, which are easily washed from farmlands into the Nisqually whenever it rains, pathogens pose a threat to the health of people living downstream.

Several organizations, including the U.S. Department of Agriculture's Natural Resource Conservation Service, Washington State University Cooperative Extension and county Conservation Districts are helping area farmers and livestock owners mend their ways. Through workshops, classes and educational materials, area farmers learn to take a more active role in protecting the Watershed. By erecting streamside fences to keep livestock from trampling banks and by properly handling excess herbicides, pesticides and fertilizers, the agricultural community does its part to keep the Nisqually River clean.

◄ Clean water and productive pasturage can coexist—providing that adequate protective measures are in place.

DELTA

DELTA

Life in the Lower Watershed

The features of the lower Watershed were sculpted about 15,000 years ago, at a time when much of southern Puget Sound lay pinned beneath the Vashon Ice Sheet. As this giant formation of frozen water and rock debris advanced and retreated, it reworked the terrain, leaving behind a sparse, flattened landscape, punctuated by potholes and underlain by coarse, gravelly soils.

Many of the Watershed's bodies of water evolved from these lifeless pools, passing through successive stages of the process called eutrophication to become inviting, biologically productive lakes. As glacial meltwater inundated the gravelly ground, it traveled to the Watershed's principal aquifer, a vast and largely uncharted underground labyrinth.

The flattened lands gradually acquired their first carpets of mosses, lichens and other primitive low-growing plants. After many centuries, grasses started to grow on these primordial plains. Trees sprouted on the perimeters, rimming the new prairies with tall Douglas fir and statuesque Garry oak. With the arrival of the Squalli-absch people, the steady encroachment of these trees onto the prairie was kept in check. Like many other Northwest tribes, the Squalli-absch managed their lands by setting fire to the prairie.

Intentional burning served several purposes. It drove elk, deer and other game into hunting range and helped give the competitive edge to perennial

◄ ◄ (PREVIOUS PAGE)

The watery world of

the Nisqually National

Wildlife Refuge.

◄ *A harvest boat works*

the outermost edge of the

delta, encompassed by a

rainbow's ethereal arc.

bunchgrasses—food for the horses acquired by the Squalli-absch through trades with tribes east of the Cascade Mountain Range.

Fire also enhanced the production of camas, a beautiful, cobalt blue flower that dotted the fire-scarred landscape each spring. Considered a delicacy by the Squalli-absch, camas bulbs were dug with a sharpened, two-pronged stick fashioned from a Pacific yew tree. Collected each spring, the bulbs were taken back to the villages, slowly steam-baked and eaten or dried for later feasts. The bulbs' taste, according to Northwest explorers Merriwether Lewis and William Clark, was sweet, like pumpkin, and with a texture that resembled an onion.

THE PASSING OF PRAIRIES

Largely because of burning, prairies may have been more abundant in the Watershed before Europeans settled in the area. According to 19th-century missionary Francis Blanchet, there were 18 prairies between the towns of Cowlitz and Nisqually. Individual prairie patches were reportedly 15 to 20 miles long and nearly as wide. The annual cycle of fire on the prairie ended with the takeover of Northwest tribal lands, making it easier for the forest to creep in from the outer fringes. Other plants invaded, led by Scot's broom, a fast-growing shrub with yellow flowers introduced to Washington in the mid-1800s. By overshadowing competitors, the aggressive Scot's broom displaces native prairie plants. It has a particularly devastating affect on water-retaining mosses and lichens that keep prairie soils moist, and thwarts most attempts made by native flora to repopulate the landscape.

Camas bulbs were staples of the Nisqually Indians. While no longer common, deep-hued camas blooms are still food for the soul.

Land practices of European settlers also hastened the demise of prairies. To the newcomers, open spaces looked perfect for livestock production. This perception proved disastrous for native grasses and herbs, many of which were soon eliminated and replaced by aggressive species, introduced from Europe and the East.

Sheep once grazed the largest of the Watershed's open spaces. Flocks belonged to the Hudson's Bay Trading Company, a British-owned enterprise that, in the early 1800s, had holdings throughout the Pacific Northwest. The company's representative, Archibald MacDonald, came to the Watershed on a scouting expedition in 1832. His mission: to find a convenient midpoint for trade between two established Hudson's Bay outposts—Fort Vancouver at the mouth of the Columbia River and Fort Langley on the Fraser River in British Columbia.

◄ (PREVIOUS PAGE) *Cleared land is soon dominated by newcomers—both native and introduced plant species, including these colorful foxgloves in bloom.*

► *Locust trees and lines of logs delineate the upland site of Fort Nisqually.*

96

"Your first objective is to observe if the Soil is suitable for cultivation and the raising of cattle; the next, the Convenience the situation affords for Shipping," read MacDonald's instructions from his Chief Factor, Dr. John McLoughlin. To MacDonald, the Nisqually River Watershed had both, so he chose this region for the new Puget Sound post. At his urging, the Hudson's Bay Company initially erected a storehouse at Sequalitchew Creek near the mouth of the Nisqually, then, after moving this structure onto the plateau overlooking the Sound, they added officers' quarters, a general store and a stockade.

Most of the Hudson's Bay Company's interest in Fort Nisqually centered around livestock production, especially after local supplies of beaver, bear, raccoon and otter skins were exhausted. A new enterprise was launched, expressly to supply other settlements with meat, milk and tallow. Sheep arrived in 1838 from Mexican ranches and company outposts in California. Although losses were large (160 of the first 800 head perished in transit), eventually about 12,000 of these animals foraged on the newly formed Puget's Sound Agricultural Company's land. Other livestock—some 3,000 head of cattle and 300 horses—also grazed on the prairie.

STEADY CHANGE

By the mid-1840s, Britain no longer held a controlling interest in the Watershed and its surrounding territories, and much of the Puget's Sound Agricultural Company's land holdings were now in the hands of American squatters. With the Treaty of 1846, the land became officially American, but

Fort Nisqually continued to function under British leadership until fair compensation could be arranged. In 1869, after four years of arbitration, the U.S. government awarded the Puget's Sound Agricultural Company $450,000 to relinquish its trade rights and claims.

Shortly after the land passed to American hands, the U.S. Army built a blockhouse at Yelm Creek to guard the livestock and their American owners from hostile Indians, thus facilitating further settlement of the prairie. Fort Nisqually's log buildings and surrounding land were purchased by Edward Huggins, former Hudson's Bay Company clerk and last resident of the outpost. Thirty years after Huggins' death, crews of the federal Works Progress Administration moved the two remaining Hudson's Bay Company buildings to Point Defiance Park in Tacoma. The restored fort remains a popular attraction to this day. In the Watershed, a pair of commemorative plaques and some locust trees grown from imported seeds are all that remain to remind us of Fort Nisqually's presence.

HOMES ON THE PRAIRIE

The Watershed's prairie habitat changed again with the completion of the Portland-to-Tacoma railway. Cutting across the grasslands, this line linked the tiny sheep-ranching and crop-farming settlement of Yelm with larger, more distant commercial markets. It was also a harbinger of further changes. In 1883, a rough road through the prairie to Longmire Springs was completed, bringing more heavy traffic and further disrupting prairie habitats.

◄ *Fort Nisqually's granary (pictured here in 1905) is considered the oldest building in Washington State and is now on display in Tacoma's Point Defiance Park.* (BACKGROUND) *Fall colors in the Watershed.*

99

The prairie was transformed again as the Yelm Irrigation Ditch project rerouted its natural freshwater supply. Completed in 1916 (after four years of hard labor) and remaining in service until the early 1950s, this project diverted water from the Nisqually River and channeled it into a long earthen ditch, where water was divvied up by area farmers. The immediate result: 5,000 acres of prairie converted to pasturage and commercial berry patches. Although no longer in service, sections of the Yelm Irrigation Ditch are still visible today, a memorial to its builders and the prairie they displaced.

Homesites replaced other prairie lands—a conversion easily accomplished on the fairly flat, treeless plain. Today, grassland ecologists believe that only a third of the Watershed's historic prairie acreage has survived. Within this acreage are a number of endangered, threatened or special plant species. Several of the prairie's vanishing plant communities are also considered unique. Without the habitats these plants and communities provide, many other living things—including the endangered Mazama pocket gopher and streaked horned lark—might also disappear.

▶ Viewed from the air, the Watershed's fog-shrouded prairies appear surrounded by ever-encroaching conifers.

▶ ▶ (FOLLOWING PAGE) A vast sea of prairie grasses greeted settlers of the lower Watershed.

PRAIRIE REMNANTS

Today the Watershed's healthiest prairie remnants are found on the troop training grounds and artillery ranges of Fort Lewis Military Reservation. Periodic shelling and the maneuvers of U.S. Army tanks and foot soldiers have proved less disruptive to the prairie environment than irrigation, home building and unrestricted foraging by cattle and sheep.

Some of the Army's activities may have unintentionally led to improved prairie conditions. Occasional brush fires, for example, on the 7,000-acre Artillery Impact Area slowed the invasion of Douglas fir and Scot's broom. At other sites, however, the Army has taken an active stance, working cooperatively with the Washington Chapter of The Nature Conservancy to hand-pull invasive weeds from its grounds. And each year, prairie ecologists for the Army supervise the planned burning of roughly 2,000 acres of prairie.

Several blossoming prairie plants—small-flowered trillium, white-top aster and several plants considered sensitive, threatened or endangered in Washington State—are today holding their own at locales like Weir Prairie (protected since the late 1940s) and 13th Division Prairie (set aside with the founding of Camp Lewis in 1917). So are Idaho fescue, a spiky, knee-high native bunchgrass, and small patches of giant camas—a taller, larger-flowered version of the Squalliabsch's prized bloom. These plants attract other colorful prairie treasures, among them the highly imperiled Mardon skipper and Edith's checkerspot butterflies. Attracted only to certain flowers, these two species of plant pollinators may play a key role in maintaining the diversity of their habitat.

Surrounding Fort Lewis' 12,000 acres of prairie are several prime examples of oak and oak-conifer stands, plus the only extensive ponderosa pine forest found west of the Cascade Mountain Range. Some of these trees are believed to be at least 250 years old. They offer food and shelter to the endangered western gray squirrel, a silver gray mammal with a long bushy tail edged with white. Dependent on such rare stands, western gray squirrels have all but vanished from other parts of the Northwest. A rough count in 1992 revealed approximately 40 squirrels within the boundaries of Fort Lewis.

Tiger swallowtail butterflies add another layer of color to the prairie pallete.

Recent studies suggest that the western grey squirrel may also be playing an important part in preserving the prairies. In addition to acorns and pine seeds, this animal's diet includes the tender cambial layer within the bark of certain young trees. By gnawing at the bark, the squirrels effectively girdle and kill young Douglas firs, keeping these conifers from encroaching on Fort Lewis' prime properties.

PUSH AND PULL OF THE TIDES

For the final three miles of the Nisqually, no canyons or hydropower installations inhibit its journey. But more subtle influences direct its flow and shape its path. As the Nisqually moves lazily across the lower Watershed's plain—a broad, flat expanse filled with reedy marsh plants and pungent estuarine mud—the water in its silt-bottomed main channel is alternately daunted and encouraged by Puget Sound's saltwater tides. Extremely low tides draw the Nisqually seaward, measurably hastening its current. Strong incoming tides act like a giant piston, actually pushing it back, raising water levels and momentarily reclaiming the Nisqually's high banks.

Throughout this tidally induced game of tug-of-war, the Nisqually dumps sand, silt, soil and organic debris on either side of its channel and at its mouth—a process called deposition. Layer upon layer of material builds up over consecutive years, and coastal waves sculpt the deposits, dramatically changing the shoreline. In the Nisqually River Watershed, the result is a fan-shaped landmass called the Nisqually River Delta. The delta's seaward edge—a

▶ Throughout the year, the delta's marshes and mudflats are visited by thousands of seabirds, shorebirds and waterfowl.

fairly featureless gray mudscape transected by dozens of small, interconnected channels—is known as the Nisqually Flats. Beyond the flats lies the Nisqually Reach, a broad shoal of mud and sand permanently covered by salt water.

In places, the Flats' silt and fine-grained sediment are 20 feet deep. Crabs, shrimp, snails, worms and other invertebrates dig tunnels and burrows or build parchment-like tubes in the first foot of the mudflat environment. A square foot of Nisqually River Delta mud may contain thousands of these cold-blooded mudflat residents. Some creatures fill even deeper niches. The geoduck, largest clam species in North America, digs down three feet, snaking its long, shotgun barrel-shaped siphon up through the muck and feeding on Puget Sound's plankton-rich soup. The fare, while microscopic, is filling: the geoduck's 6-to-10-pound body is so large that it cannot be fully withdrawn into the clam's shell.

Directly inland from the mudflat lies another important Nisqually Delta habitat—the salt marsh. The partially submerged turf is dominated by pickle-weed, tufted hairgrass, slough sedge and other salt-tolerant plants. Most of these plants, called halophytes, grow in tangled masses, their knotted roots aiding in the slow but steady build-up of delta land by capturing silt and soil particles. The tangled root masses also support communities of animals—everything from tiny pill bugs to larger rodents and birds.

► *Within the Nisqually National Wildlife Refuge, rushes and other wetland plants grow in dense, undisturbed mats.*

DIKED AND DRAINED

A relic from pioneer days, the heaped earth of an historic dike separates seawater from semi-dry land on the delta. Many European settlers viewed wetlands

as worthless swamps, breeding grounds for disease and vermin. Others recognized wetlands as valuable real estate, with flat, treeless terrain and fertile soils waiting to be turned by the plow. From either perspective, wetlands throughout Washington have suffered from dredging, draining and diking operations. More than half of Washington's original wetlands have been destroyed or seriously altered. Only after their disappearance do we recognize their value—as nurseries for juvenile salmon and other aquatic organisms, anchors against shoreline erosion, filters for particles and pollutants and barriers to flooding.

The Nisqually River Watershed's largest wetland was threatened in 1904, when Seattle attorney Alson Brown bought 1,500 acres on the Nisqually River Delta and another 850 acres on the hillside overlooking McAllister Creek. Intent on farming this land, he built four miles of earthen dikes on his property's

◄ ◄ *Haying of grasslands within the refuge makes the land more productive, nurturing wintering waterfowl.*

◄ *Today in the lower Watershed, farms and wetlands coexist.*

▲ *Throughout Puget Sound, people once shared the bounty from Brown Farm.*

▶ *Oysters from the delta are enjoyed by diners near and far.*

eastern, western and northern borders, effectively walling off his land from the sea. Wooden one-way gates in the north dike kept out salt water from Puget Sound but let fresh water from upland sources percolate through the property, removing any residual salt with it. After three years of leaching, the salt content of his soil had subsided enough for crops to grow on the reclaimed land.

By 1914, Brown Farm supplied fresh farm products for the growing Puget Sound market. He maintained a herd of 300 milk cows, tended 1,200 hogs, kept 4,000 egg-laying hens and cultivated several swarms of honeybees.

"Continuous motion, just like the Ford's assembly plant," recalled farm foreman Charles Rough, remembering the days when he and another worker shoveled food for 500 hogs kept in 50 pens on either side of one central aisle. The animals' droppings passed through a pipe and discharged into the Nisqually River—a straightforward solution to waste disposal but highly detrimental to water quality downstream. Today, under the federal Clean Water Act, farmers must get special permits, similar to those for industrial discharges, before allowing contaminated water from large confined animal feedlots to enter rivers or streams.

WATERY PATCHWORK OF PLACES

When the Brown Farm acreage went up for sale in the 1960s, both the cities of Seattle and Tacoma proposed industrial uses for the land. However, local efforts to preserve the delta and its essential feeding and breeding grounds for wildlife took precedence. Hundreds of species of fish, amphibians, birds and

mammals depend on the delta for sustenance. As a critical link in the Pacific Flyway, a migratory route that originates on the west coast of Central America and ends in Alaska, the delta offers vital resting and feeding opportunities for long-distance travelers such as dunlins and short-billed dowitchers. During their stopovers at the delta, these two sandpiper species feed intensively. Using their sensitive tubular bills to probe the mud for morsels, they sometimes double their weight in less than two weeks. But any fat the birds gain is quickly lost after they take off, as these drab-feathered visitors sometimes fly thousands of miles before stopping again.

Each spring and fall, the delta's freshwater marshes also become stopovers for Canada goose and other migratory waterfowl, and serve as feeding, resting and brood-rearing places. Transition areas—lands once flooded with salt water but now diked to prevent spillover from Puget Sound—are foraging grounds for American wigeon, mallard and pintail. Several threatened and endangered inhabitants of Washington State also reside here. The six- to eight-inch-long northwestern pond turtle was once found in watershed ponds, lakes and marshes throughout the southern Puget Sound region. Particularly susceptible to habitat loss and degradation, this docile creature needs large, undisturbed tracts of aquatic land to survive.

◄ Hungry birds wait for low tide at the Nisqually Flats.

Many other animals, some threatened and endangered, rely on wetland habitats for rearing young. A pair of bald eagle nests occupies the bluffs above McAllister Creek, their builders listed as threatened in Washington and endangered in 43 other states. Not far from these nests, endangered peregrine falcons hunt. Nearby, a colony of about 60 breeding pairs of great blue herons shares the bluffs.

The nests of herons are small and unsophisticated compared to those of other large birds, like ospreys or eagles. Hastily gathered piles of twigs, branches and tree limbs, these nests are stacked closely together, as many as four or five nests to a tree. In each nest are three to six bluish-green ovals, each about the size of a chicken's egg. When the eggs hatch, the devoted parents make daily trips between coastal feeding grounds and their nests. It is not uncommon for great blue herons to cover a distance of 10 miles each way, just to feed their young.

SAFE HAVEN

Recognizing the importance of the Nisqually River Delta—one of the largest wetland systems in Washington—the state Department of Game bought 620 acres in 1969, effectively limiting development in these key habitats. Two years later, the area outside the farm dike was designated as a National Natural Landmark. Enhancing both preservation efforts, the U.S. Fish and Wildlife Service purchased the land for the Nisqually National Wildlife Refuge in 1974.

Lands within the refuge have been set aside specifically for wildlife, with public access restricted to nature trails and designated fishing and wildlife viewing areas. As of 1994, the refuge contained approximately 4,100 acres, about two-thirds of which are under federal ownership. Despite the heavy visitor traffic, estimated at 70,000 people per year, the close proximity of well-traveled Interstate 5 and the steady development of surrounding acreage, the delta retains its ecological integrity, remaining a safe haven for Watershed wildlife

With a swift jab of its beak, a great blue heron snares a fish or a frog, then resumes its routine of wading, watching and waiting.

◀ *Located on the west side of the delta, the Nisqually Reach Nature Center offers environmental education opportunities.*

while providing important aesthetic, educational and passive recreational opportunities for people.

WE ALL LIVE DOWNSTREAM

The water that pours from the Nisqually River into Puget Sound is laden with everything from grains of sand and soil particles to chemicals, lawn clippings and, after strong storms, tree limbs and other large debris—carried by the currents from all corners of the Watershed. Looking at these riverborne souvenirs, it's easy to understand how every action, human and natural, is interrelated. Now at sea, this water eventually returns to the Nisqually Watershed, some as moisture wrung from the clouds atop Mount Rainier. The water's journey through the river and its watershed begins anew.

Many forces will shape this long, winding path to the sea. At almost every mile of this trip, there are signs of the Nisqually River Watershed's overall wellness. Some are attributed to the cooperative efforts of groups, others to individual actions and still others, perhaps, to nature's ability to heal itself over time.

◄ Sunset at the refuge.

Success stories like Mount Rainier National Park, Vail Tree Farm and the Nisqually National Wildlife Refuge are scattered throughout the Watershed. Possibly the greatest success in recent years has been conceptual—the recognition by Watershed stakeholders that the Nisqually River's history, culture, environment and economy are all tightly entwined. These four components of Watershed life must be carefully considered together, for any sound, balanced management decisions to be made.

An Agenda to Educate and Involve

WATERSHED EDUCATION has been a high priority of the Nisqually River Council. By forming partnerships with federal, state and local governments, school districts, the Nisqually Indian Tribe and area businesses, the Council encourages and funds an array of projects to inform and involve people in conserving the Nisqually.

Especially well received has been Nisqually River Project GREEN (Global Rivers Environmental Education Network), the local branch of an international water quality monitoring grid that recruits students to help safeguard the health of the Nisqually River and its tributaries. At several sites in the Watershed, students collect chemical, physical and biological data each month, contributing valuable information to salmon fisheries programs of the Nisqually Indian Tribe. Via an electronic bulletin board, this information is shared with other Project GREEN participants throughout Washington and in over 130 countries around the world.

Equally popular among young people are special school programs about the Watershed. Coordinated by the Yelm School District and funded by federal, state and private organizations, the Nisqually River Education Project reaches nearly 2,000 students a year with site-specific curriculum materials, field trips and opportunities for hands-on water quality monitoring. Teacher

118

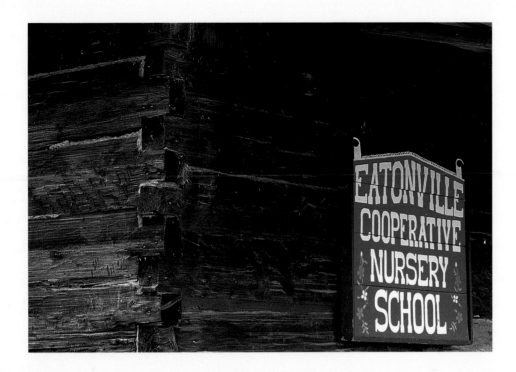

training is also an important part of this multi-faceted endeavor, empowering educators to bring the word about Watershed stewardship back to their classrooms.

To foster a sense of place among area residents and their guests, the Council has worked with the state to erect informational signs wherever major roads cross the Watershed. It also publishes the *Nisqually River Notes* newsletter and created a Nisqually Basin Watch to encourage people to report illegal burning, dumping or poaching.

The Nisqually River Interpretive Center Foundation was formed in 1992 to act on the Nisqually River Management Plan's recommendation for a Watershed-wide network of interpretive and educational sites. With start-up funds from the State Legislature, augmented by private donations, this nonprofit group drafted plans for a major facility at which Watershed residents, visitors and students can view and interact with exhibits of regional nature, economics, history and culture.

All of these education and involvement activities will promote the thoughtful management of the Nisqually River, impart awareness of the value of cooperation and foster a stewardship ethic among residents of the Watershed.

◄ ◄ *Youth plays an important part.*

◄ *Fresh face on Eatonville's first schoolhouse.*

▶ *All of us have a stake in our watersheds' health.*

▶ ▶ *From glacier to delta: we all live downstream.*

Everyone has a part to play in watershed stewardship. To some, this means conserving water, using environmentally friendly lawn and garden products or composting and recycling their reusable waste. To others, it is learning to become an active voice in their watershed, participating in local planning efforts or getting involved in the process of reviewing land use permits and crafting new laws. Still others may choose to volunteer with a service organization, pooling their skills with watershed residents to replant streambanks, clean roadside ditches or help restore salmon runs.

By working together, we can help conserve the world's watersheds—passing the gifts of productive land and clean water to future generations.

Publishers' Notes

THE NISQUALLY RIVER INTERPRETIVE CENTER FOUNDATION was founded as a nonprofit organization in 1993. Its mission is *"fostering a stewardship ethic by providing interpretive and educational opportunities that emphasize the system of natural, cultural, historic and economic resources of the Nisqually River Basin."*

An important element of the Foundation's education and interpretation program is the development of a central facility to serve as a study center and repository for cultural and natural history information. The facility will provide contact and referral services to other interpretive and educational facilities in the watershed, serve as a center for the study and development of emerging cultural, historic and economic resource policy, and create a place that, through its design, presents and reflects the native landscape of the watershed.

To learn more about the Nisqually River Interpretive Center or to become involved in the Foundation's programs, write the Nisqually River Interpretive Center Foundation, P.O. Box 759, Yelm, Washington 98597.

❖

THE MOUNTAINEERS, founded in 1906, is a nonprofit outdoor activity and conservation club, whose mission is *"to explore, study, preserve and enjoy the natural beauty of the outdoors. . .".* Based in Seattle, Washington, the club is now the third-largest such organization in the United States, with 15,000 members and four branches throughout Washington State.

◄ *Alders near Elbe.*

The Mountaineers Books, an active nonprofit publishing program of the club, produces guidebooks, instructional texts, historical works, natural history guides and works on environmental conservation. All books produced by The Mountaineers are aimed at fulfilling the club's mission.

To participate in the organized outdoor activities of the club's programs or receive a catalog of more than 300 outdoor books, write or call The Mountaineers Books, 1011 Southwest Klickitat Way, Suite 107, Seattle, Washington 98134; 1-800-553-4453.

Additional Information

BOOKS

Cecelia Svinth Carpenter, *Where the Waters Begin: the Traditional Nisqually Indian History of Mount Rainier* (Seattle: Northwest Interpretive Association, 1994)

Gene Allen Nadeau, *Highway to Paradise: a Pictorial History of the Roadway to Mount Rainier* (Puyallup, Wa.: Valley Press, 1983)

Arthur D. Martinson, *Wilderness Above the Sound: the Story of Mount Rainier National Park* (Niwott, Co.: Roberts Rhinehart Publishers, 1994)

William H. Moir, *Forests of Mount Rainier* (Seattle: Pacific Northwest National Parks and Forests Association, 1989)

EDUCATOR'S GUIDE

Where the River Begins: the Nisqually River of Mount Rainier National Park; The Living River: the Nisqually River Basin; and Where the River Meets the Forest: the University of Washington Pack Experimental Forest (produced through the Nisqually River Education Project, 1994 and 1995)

ORGANIZATIONS:

Nisqually River Interpretive Center Foundation, P.O. Box 759, Yelm WA 98597

Mount Rainier National Park, Tahoma Woods - Star Route, Ashford WA 98304

Nisqually Delta Association, P.O. Box 7444, Olympia WA 98507

Nisqually Indian Tribe, 4820 She-Nah-Num Drive SE, Olympia WA 98513

Nisqually National Wildlife Refuge, 100 Brown Farm Road NE, Olympia WA 98506

Nisqually Reach Nature Center, 4949 D'Milluhr Road NE, Olympia WA 98506

Nisqually River Basin Land Trust, P.O. Box 1148, Yelm WA 98597

Nisqually River Council and its Citizens Advisory Committee, P.O. Box 1076, Yelm WA 98597

Nisqually River Education Project, P.O. Box 476, Yelm WA 98597

Northwest Trek, 11610 Trek Drive E, Eatonville WA 98328

Pioneer Farm Museum, P.O. Box 1520, Eatonville WA 98328

University of Washington Pack Experimental Forest, 9010 453rd Street E, Eatonville WA 98328

◄ *Day's end in the delta.*

Index

Page numbers of photographs indicated in **bold.**

Alder (town) 60
Alder Lake **5,** 60, **61**
Ashford 47, 48, **48,** 51, 54, 75
Awards (watershed management) 32

Bald Hill National Area Preserve 81
Birds 43, 45, 79, **105, 112,** 113
 Dippers 43
 Eagles, bald 113
 Falcons, peregrine 113
 Herons, great blue 113, 114
 Lark, streaked horned 100
 Sandpipers 113
 Swallows 67, 68
Blanchet, Father Francis 94
Bridge **18, 29, 30, 37, 66**
Brown, Alson 109, **109**
Butterflies 103

Camas 94, 103
Champion International 75
Citizens Advisory Committee 32, 125
Clean Water Act 111
Confluence
 Nisqually and Mashel 68
 Nisqually and Ohop 82
Cooperation 23, 28, 31, 32, 82, 85, 89, 103, 117-119

Dams 56-63
 Alder **57, 58,** 59, 63
 Centralia 60, 62, 63
 LaGrande 56, **57,** 59, 63
Dike (Brown's farm) 106-113

Eatonville **74,** 75, **76,** 85, **119**

Education 72, 89, 116-119, **118,** 123
Elbe 43, 56
Elk 25, 45
Endangered species 100, 103, 113
Eutrophication 93

Farming 28, 82, 85, 86, **87, 88,** 89, 100, **108, 109,** 111
Fiander, Robert 82
Fire 80, 93, 94, 103
Fish ladder 62
Fishing, see also Hatcheries, **35,** 45, 62, 63, **63, 120**
Forests and forestry 28, **41,** 75, 76, **78,** 79, **79,** 80, 81, 93, **98,** 103, 104, **122**
Fort Lewis Military Reservation 63, 72, 75, 99, 100, 103, 104
Fort Nisqually **96,** 97, **98,** 99
Friends of the Earth 32
Frog, red-legged 72

Glaciers, see also Nisqually Glacier, 20, 23, 38, 53, 54, 55

Habitat 26, 34, 43, 60, 76, 81, 99, 100, 103, 113, 114
Hatcheries 63
Hellyer, David and Connie 80, 81
Homesteaders, see Pioneers
Hudson's Bay Trading Company 96, 97, 99
Huggins, Edward 99
Hydrologic cycle **17,** 25, 26, **34,** 117
Hydropower 28, 55-63, 104

Indian Henry (So-to-lic) 46, 72, 82
Indian Sam (Isalahah) **44**
Indian Wars (Puget Sound) 70, 82

Kjelstad, Matteus 82, 85

Legislation 31, 34, 40, 111, 119
Leschi, Chief 70, **70**
Lewis and Clark 94
Livestock **74,** 82, 86, 89, 96-100, 111
Logging 40, 48, **50, 51,** 68, 75, 76, **77,** 79
Longmire 46, **46,** 47, **47,** 51, 55, 99

Mashel Prairie 85
Mashel River **30,** 43, 68, **69,** 72, 76, 79, 82, 89
Massacre 70
Maxon, Captain Hamilton 70
McAllister Creek **109,** 113
McDonald, Archibald 96, 97
McKenna **78**
Me-schal 68, 70
Mount Rainier **6, 17,** 19, 39, 47, 54, 55, **71,** 117, **121**
Mount Rainier National Park **17, 18, 22, 38,** 40, 46, **46,** 47, **47,** 51, 53, 56, 80, 117, 125
Murray Pacific 75

Narada Falls **18, 37**
National (town) 47, 48, **50,** 51, 75
Native history 11-13, 43, 45, 68-72, 82-85
Natural Heritage Program 81
Nature Conservancy 81, 103
Nisqually Flats **33,** 106, **112**
Nisqually Glacier 20, **20,** 23, 39, 54, **54,** 55, **55**
Nisqually Indian Tribe 11-13, 31, 43, 45, 62, 63, 68, 70, 75, 82, 118, 125
Nisqually National Wildlife Refuge **91, 107,** 114, **116,** 117, 125

Index

Nisqually Reach 106
Nisqually Reach Nature Center
115, 125
Nisqually Reservation 72
Nisqually River
Lower **27,** 91-117, **91**
Middle **3, 10, 24,** 67-89
Origins **21,** 23-26, 39, 40
Upper **22, 38, 42,** 38-66
Nisqually River Basin Land Trust
81, 125
Nisqually River Basin Watch 119
Nisqually River Council 32, 118,
125
Nisqually River Delta 60, **91, 92,**
104-114, **105, 121, 124,** 125
Nisqually River Education Project
118, 125
Nisqually River Interpretive
Center Foundation 119, 125
Nisqually River Management Plan
31, 32
Nisqually River Notes newsletter
119
Nisqually River Project GREEN 118
Nisqually River Task Force 31
Nisqually (town) 94
Northern Pacific Railroad
Company 80
Northwest Trek Wildlife Park 80,
80, 81, 125

Ohop Creek **65, 66,** 82, 85, 89
Ohop Valley 82, **83,** 85, 86

Pacific Flyway 113
Pacific National Lumber
Company 48, **51**
Pack, Charles Lathrop 81, **81**
Pack Experimental Forest 79, **79,**
80, 81, 125
Paradise River **18**
People, effects of 26, 28, **29,** 40,
76, 89, 99, 107, 109, 117, 120
Peterson, Torger 82
Pioneer Farm Museum **84,** 85,
85, 86, 125

Pioneers 45-56, 72, 82, 85, 86, **86**
Plants 94, **95,** 100, 103, 106
Plum Creek 75
Point Defiance Park **98,** 99
Portland to Tacoma Railway 99
Prairie 93-104, **101, 102**
Preservation 80, 81, 103, 111, 114
President's Council on
Sustainability 32
Protection 26-31, 40, 80, 81, 89
Puget Sound 23, 26, 104, 106, 117
Puget Sound Agricultural
Company 97, 99
Puget Sound Water Quality
Management Plan 28-31

Quaymuth 70

Railroads **29,** 48, **49,** 51, 56, 60,
75, 80, **83,** 99
Recreation **35, 38,** 46, **46,** 47, 53,
55, 60, 80, 81, 85, 86, 99,
114, 117, 123
Reflection Lake **44**
Roadways **49,** 51, **51, 52,** 53, **53,**
56, 99
Roy 43

Salmon 23, 28, 45, 62, 63, **63,**
72, 76, 85, 109, 118
Salt marsh 106
Schools 72, 74, **119**
Scot's broom 94, 103
Sheep 96, 97, 100
Shellfish 26, 45, 106, **110**
Squalli-absch 43, 45, 46, 93, 94,
103
Squirrel, western grey 103, 104
St. Paul & Tacoma Lumber
Company **49,** 80, **83**
Stakeholders 31-35, **120**
Steelhead 62
Stevens, Isaac 70
Stewardship 23, 31, 40, 80, 119,
120

Ta-co-bet **71**

Tacoma Eastern Railway 48, 75
Tacoma City Light (formerly
Tacoma Light Department)
55, 56, 59, **59,** 60
Taft, President William Howard 56
Threatened species, see
Endangered species
Thompson Creek 60
Tides 104
Treaty of 1846 97
Tule Lake **8-9, 73**

Uniqueness of watershed 23, 31,
100, 103, 114, **121**
U.S. Army at Fort Lewis, see Fort
Lewis Military Reservation
U.S. Department of Agriculture 89
U.S. Fish and Wildlife Service 63,
114
U.S. Forest Service 75
University of Washington 75, 81
(see also Pack Experimental
Forest)

Vail Tree Farm 79, 117
Van Eaton, Thomas C. 72, 74
Vashon Ice Sheet 93

Washington Department of
Fisheries 62, 63
Washington Department of Game
62, 114
Washington Department of
Natural Resources 51, 75, 81
Washington State University
Cooperative Extension 89
Water quality **23,** 26, 28, 31, **88,**
89, 111, 118
Watershed (definition) 25, 34
Wetlands 106-111, **107, 109**
Weyerhaeuser Company 51, 75,
79, 81
Wonderland Trail **38**
World Resources Institute 32

Yelm 43, 99
Yelm Irrigation Ditch Project 100